STEM Makerspace Projects

MAKERSPACE PROJECTS FOR UNDERSTANDING PLANT SCIENCE

RACHAEL MORLOCK

PowerKiDS press
New York

Published in 2021 by The Rosen Publishing Group, Inc.
29 East 21st Street, New York, NY 10010

Copyright © 2021 by The Rosen Publishing Group, Inc.

All rights reserved. No part of this book may be reproduced in any form without permission in writing from the publisher, except by a reviewer.

First Edition

Editor: Danielle Haynes
Book Design: Reann Nye

Photo Credits: Series art (background) ShutterStockStudio/Shutterstock.com; cover DUSAN ZIDAR/Shutterstock.com; p. 5 SDI Productions/E+/Getty Images; p. 7 Chutima Chaochaiya/Shutterstock.com; p. 9 Science & Society Picture Library/Getty Images; p. 11 Olivkairishka/Shutterstock.com; p. 12 Lucky-photographer/Shutterstock.com; p. 13 yuris/Shutterstock.com; p. 14 R7's Photo/Shutterstock.com; p. 15 VectorMine/Shutterstock.com; p. 19 Brian A Jackson/Shutterstock.com; p. 21 herain kanthatham/Shutterstock.com; p. 25 Rafal Miszkurka/Shutterstock.com; p. 29 Hero Images/Getty Images.

Library of Congress Cataloging-in-Publication Data

Names: Morlock, Rachael, author.
Title: Makerspace projects for understanding plant science / Rachael Morlock.
Description: New York : PowerKids Press, [2021] | Series: STEM makerspace projects | Includes index.
Identifiers: LCCN 2019032638 | ISBN 9781725311824 (paperback) | ISBN 9781725311848 (library binding) | ISBN 9781725311831 (6 pack) | ISBN 9781725311855 (ebook)
Subjects: LCSH: Botany projects–Juvenile literature.
Classification: LCC QK52.6 .M67 2021 | DDC 580.72/4-dc23
LC record available at https://lccn.loc.gov/2019032638

Manufactured in the United States of America

CPSIA Compliance Information: Batch #CSPK20. For Further Information contact Rosen Publishing, New York, New York at 1-800-237-9932.

CONTENTS

HANDS-ON LEARNING 4
THE GREEN WORLD 6
DISCOVERING PLANTS 8
CLASSIFYING PLANTS 10
ROOTS, STEMS, AND LEAVES 12
SOLAR POWERED 14
PROJECT 1: MAKE AN HERB GARDEN 16
A PLANT'S LIFE 18
MAKING SEEDS 20
PROJECT 2: BUILD A BUG HOTEL 22
SUPPORTING SURVIVAL 24
PROJECT 3: MAKE A WORM COMPOSTING BIN 26
HIGH-TECH PLANTS 28
GOING GREEN 30
GLOSSARY 31
INDEX 32
WEBSITES 32

HANDS-ON LEARNING

Sometimes the best way to learn is to take an active part in exploring, discovering, or creating something. A makerspace is a workshop for hands-on learning and making things. It's a place where people of all ages can come together, share tools and materials, and try something new.

Makerspaces come in different forms. You might find them in libraries, schools, or community centers. But it's not what a makerspace looks like that matters, it's what happens there. A makerspace can be anywhere people come together to work, learn, make, and share. You could even create a makerspace in your home. It's a place where you can learn with others about new ideas or **technology**. Then you can understand how they work by trying them out yourself.

MAKER MAGIC

Some makerspaces have special technology that people can learn to use, including 3-D printers, sewing machines, or woodworking tools. Other makerspaces offer simple materials for crafting and making.

What kind of makerspace can you use or create to make the projects in this book?

THE GREEN WORLD

Makerspaces can help you learn about big ideas in science, technology, and art. They give you an up-close look at the processes that make these ideas work. The makerspace projects in this book can help you understand plant science through firsthand experiences and experiments you can do yourself.

Plant science, also called botany, is the study of the life of plants. It explores and explains the structure of plants and the way they grow and reproduce. It also divides them into **categories** based on their form and function. Plants play an important part in the ecosystems they belong to, and plant science looks at their role in the natural world. Plant science also studies the ways plants can be used to feed people and animals, treat diseases, and contribute to technology.

What Is a Plant?

People used to categorize all living things as either plants or animals. Today, scientists know that there are many living things—such as algae, fungi, and bacteria—that aren't quite animals or plants. Now, they **define** plants as organisms that are made up of more than one cell, can make their own food, and generally have roots, stems, and leaves.

Plant scientists can work in laboratories, greenhouses, or out in nature in order to study the life of plants.

DISCOVERING PLANTS

Modern science and technology have made advanced knowledge possible, but even early humans understood the importance of plants for life on Earth. Plants provided the basic materials early humans used for food, tools, shelter, medicine, and more. As human civilization progressed, so did the study of plants and their uses.

In 300 BC, an ancient Greek philosopher named Theophrastus founded the field of plant science by writing essays about plants. A few centuries later, another Greek writer, named Pedanius Dioscorides, **described** and illustrated 600 plants. In the years that followed, people continued to study plants and the ways they could be used. The study of plants became more detailed and exact after the invention of the microscope in 1590. Scientists were then able to carefully observe the tiniest parts of plants.

Robert Hooke, a 17th-century scientist, built a special microscope to help him study plant cells.

CLASSIFYING PLANTS

Theophrastus categorized plants simply as trees, shrubs, or herbs, and Dioscorides sorted them based on their uses. As the study of plants became more scientific, so did methods for organizing them. In 1753, a scientist named Carl Linnaeus published his system for **classifying** plants. This system is still used today with few changes.

Linnaeus described 6,000 plants in his book *Species Plantarum*. He labeled each plant with a two-part name. Linnaeus called this system "binomial nomenclature." The first name identifies the **genus** of the plant. The second name identifies the species. Linnaeus's new system provided rules for naming plants and gave scientists a shared language for talking about them. It also helped them track the ways plants **evolved** over time to create new species.

> The scientific name for a common sunflower using binomial nomenclature is *Helianthus annuus*. *Helianthus* is the genus, and *annuus* is the species.

ROOTS, STEMS, AND LEAVES

Most plants have roots, stems, and leaves. Roots anchor a plant and give it **access** to water and **nutrients** in the soil. There are two main kinds of roots: taproots and **fibrous** roots. Vegetables such as carrots and beets are taproots with one main root. Other plants have many fibrous roots extending into the soil.

MAKER MAGIC

Trees have roots, stems, and leaves, but the stem of a tree is called its trunk. The world's largest tree is the giant sequoia. The tallest sequoia is 316 feet (96.3 m) tall!

Some plants, such as this red pepper plant, also have flowers and fruits.

While roots usually grow downward, most stems grow upward. A response called phototropism guides the stem to grow toward light. Although plants can't move on their own, they can grow toward the sunlight or water they need to survive.

Leaves come in different shapes, sizes, shades, and arrangements. They can be useful in identifying plants since their appearance varies from species to species. Leaves are usually either simple, with one blade, or compound, with many.

SOLAR POWERED

Most plants make their own food using energy from the sun. This process is called photosynthesis. "Photo" means light and "synthesis" means to create something. Plant leaves contain a green **pigment** called chlorophyll, which can absorb sunlight. Photosynthesis begins when plants catch and store energy from the sun.

MAKER MAGIC

Plants aren't the only living things to create food through photosynthesis. Bacteria and algae do, too! In the process, these things give off the oxygen that all animals need to breathe.

"Photo" means light and "synthesis" means the process of making something. Plants use light, water, and carbon dioxide to make their own food.

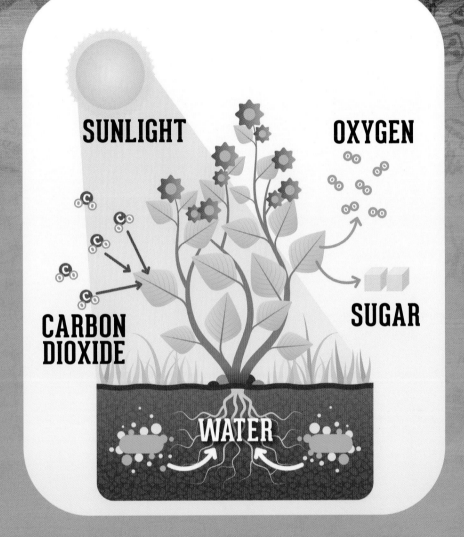

For the process to work, there also must be water and carbon dioxide in the plant's leaves. Water comes from the soil. It travels up a plant's roots and stem and into its leaves. Carbon dioxide comes from the air. It enters tiny holes in the leaves. When carbon dioxide, water, and energy come together in a plant's leaves, they create oxygen and sugars. Plants release oxygen from their leaves and use the sugars for food.

PROJECT 1: MAKE AN HERB GARDEN

With healthy soil, water, air, and sunlight, plants have everything they need to grow. You can see this in action by making your own self-watering herb garden. Can you identify the parts of your plant?

WHAT YOU NEED

- 2-liter plastic bottle
- cotton string, 32 inches long
- soil
- scissors
- ruler
- herb seed packet (such as rosemary, dill, mint, oregano, or basil)

WHAT YOU WILL DO

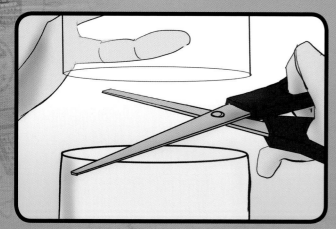

STEP 1:
Use scissors to cut the plastic bottle in half.

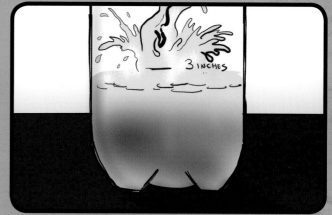

STEP 2:
Fill the bottom half of the bottle with about three inches of water.

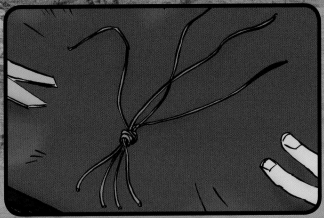

STEP 3:
Cut four eight-inch pieces of cotton string. Line up the four pieces. Tie the strings together in a knot about three inches down from one end.

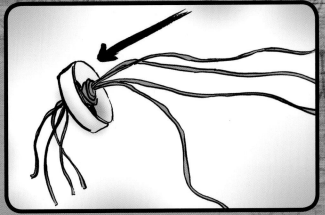

STEP 4:
With an adult's help, use sharp scissors to cut a hole in the middle of the bottle cap wide enough for all the strings (but not the knot) to fit through. Thread the short ends of the cotton strings through the hole so that they come out the top.

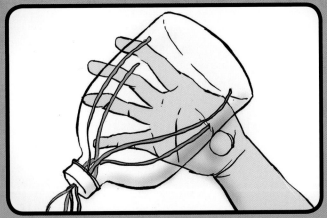

STEP 5:
Screw the lid on the top half of the bottle with the long ends of the string inside the bottle.

STEP 6:
Turn the top half of the bottle upside down and fit it inside the bottom half of the bottle, with the short ends of the strings dipping into the water.

STEP 7:
Add soil to the top half of the bottle, spreading out the cotton strings throughout the soil as you go. Fill the bottle until the soil is about an inch from the top of the container.

STEP 8:
Follow the directions on your seed packet for planting the seeds in the soil. Place the bottle planter in the sun and watch your plant grow! The cotton strings will draw water up into the soil. You can add more water if the reserve starts to get low.

17

A PLANT'S LIFE

If you watch your herb garden grow, you'll observe the early stages of a flowering plant's life. Like your herbs, most plants grow from seeds. When a seed is in a damp, dark environment at the right temperature, it begins to absorb water. It swells until its hard, outer case breaks open. This is called germination.

Next, a root emerges and grows downward, and a shoot grows upward. The shoot becomes a stem. Leaves grow, open up, and begin to use photosynthesis. When the conditions are right, the plant produces flowers.

The life cycle continues when plants use their flowers to form seeds. Seeds can come in many shapes and sizes. Their special designs help them spread out to places where they can take root and grow.

Spreading Seeds

Certain seeds fall from their parent plant and take root on the spot. Others spread far and wide using different methods. All fruits are containers for seeds. They might travel long distances when animals eat fruit and then pass the seeds as waste. Animals can also transport seeds such as burrs to a new place when the seeds become caught on their fur. Other seeds float on the water or are blown on the wind.

A seed can be as light as a bit of dandelion fluff that floats on the breeze or as large as a coconut that bobs across the ocean.

MAKING SEEDS

Pollination allows flowering plants to make seeds. This involves a flower's stamen and anthers, which are male, and pistil and stigma, which are female. Pollen from the anthers must reach the stigma in order to pollinate. Sometimes, plants pollinate on their own. Sometimes, they rely on wind, water, animals, or insects to help.

When a pollinator, such as a bee, lands on a flower, pollen attaches to its body. If the bee lands on another flower from the same species, the pollen can transfer to its stigma. A pollen tube grows down into the pistil. The plants' **genetic** material joins together into new cells. A seed forms around these new cells. This seed has everything it needs to later become a plant.

The Buzz About Bees

Bees pollinate flowers as they collect pollen or sip nectar. There are more than 4,000 bee species in the United States. Some, such as honeybees, live together in hives. Others are called solitary bees and live alone. Mason bees are solitary and nest in tunnels. They protect their eggs by making mud walls. Leaf-cutting bees also nest in tunnels, but they make a barrier out of leaves to keep their eggs safe.

Bees, butterflies, moths, beetles, bats, and birds are all pollinators. They're attracted to the bright colors or sweet smells of flowers.

PROJECT 2: BUILD A BUG HOTEL

Many insects help pollinate plants. You can encourage insects to visit your garden by building a bug hotel. Follow these instructions to make a structure that could attract leaf-cutting or mason bees, ladybugs, or other insects.

WHAT YOU NEED

- 2-liter plastic bottle
- 1 yard of strong twine
- modeling clay
- bamboo or hollow reeds and stems
- sandpaper
- scissors
- garden clippers
- a ruler

WHAT YOU WILL DO

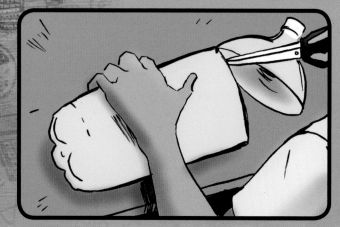

STEP 1:
Use scissors to cut both ends off the plastic bottle. This should give you a tube-shaped piece of plastic.

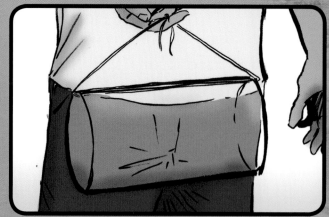

STEP 2:
Thread one end of the twine through the bottle. Bring both ends together outside the bottle and tie a knot. You will use this twine to hang the hotel once it's finished.

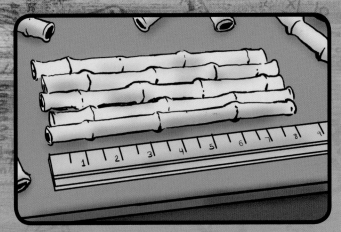

STEP 3:
Have an adult use garden clippers to trim the hollow stems. They should be about 8 inches (20 cm) long, a little shorter than the tube of plastic.

STEP 4:
Carefully sand the clipped edges of the bamboo so that there are no sharp edges that could harm an insect.

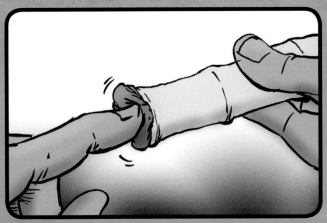

STEP 5:
Use modeling clay to create a plug in one side of each hollow piece.

STEP 6:
Tightly pack the hollow stems together inside the bottle. Make sure the hollow stems are shorter than the bottle so the plastic creates an overhang on each end to protect the stems from rain.

STEP 7:
Choose a place to hang your bug hotel. It should be at least three feet above the ground in a sunny and dry spot facing south or southeast.

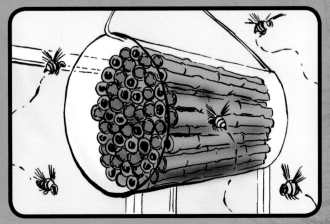

STEP 8:
Bees may use the hotel in the summer months to lay eggs. In the winter, move the hotel to a dry shed or garage. Just before the weather gets warmer, place the hotel outside again so the new insects can emerge from their nests.

SUPPORTING SURVIVAL

The life processes plants use make it possible for animals to survive as well. Photosynthesis provides the oxygen animals need to breathe. In addition to making the atmosphere safe for animals, plants also provide **habitats**. They offer shelter, protection, and a source of food to other living things. All animals, including humans, rely on plants (or other animals that eat plants) for food.

Plants play an important role in maintaining Earth's natural cycles. They contribute to the water cycle by drawing water up from the soil and releasing it into the atmosphere through a process called transpiration. Plants also keep soil healthy and intact. Their roots hold soil together, and dead plants add important nutrients to the soil as they break down. These nutrients help new plants grow.

MAKER MAGIC

Every food chain begins with plants. Plants are called producers because they make their own food. All living things on Earth need plants in order to survive, whether or not they eat plants themselves.

The nutrients in dead leaves and plants are recycled. Decomposers, such as earthworms, fungi, and bacteria, help this process by eating or breaking down dead plants.

PROJECT 3: MAKE A WORM COMPOSTING BIN

Worms are called decomposers because they eat dead plants and then return their nutrients to the soil as waste called casts. You can help worms make healthy soil with a worm compost bin. Feed your worms vegetable scraps, and they will make rich soil for your garden! Ask an adult to help you with this project.

WHAT YOU NEED

- two plastic tubs that nest together
- cardboard
- scissors
- drill
- a piece of fabric
- rocks
- shredded newspaper
- live red wriggler worms
- spray bottle
- soil
- vegetable scraps

WHAT YOU WILL DO

STEP 1:
Place the inner plastic tub upside down on a large piece of cardboard and trace the outline of its rim. Use scissors to trim the cardboard so that it will fit snugly into the top of the tub and create a lid. You can leave a small tab to make it easy to lift the lid.

STEP 2:
Have an adult drill several small holes in the bottom of the inner tub.

STEP 3:
Line the inside of the bottom of the inner tub with a piece of fabric. This will keep worms from escaping through the holes.

STEP 4:
Arrange a few rocks into the bottom of the outer tub. Place the inner tub inside it. The rocks should create a gap in the bottom for greater air flow.

STEP 5:
Place a layer of shredded newspaper and a small amount of soil in the bottom of the inner tub. Use the spray bottle to wet it.

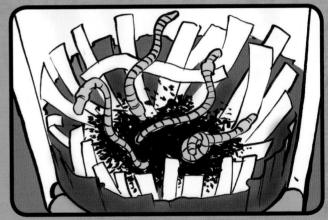

STEP 6:
Add a handful of red wriggler worms—about 1 pound of worms for every square foot of surface area inside your tub—to the tub and give them a few vegetable scraps to eat. You can also feed the worms crushed eggshells. Place the cardboard lid over the tub and store it in a dark, cool place.

STEP 7:
Wait until the worms have finished their scraps before adding more. You can occasionally add more shredded newspaper sprayed with water to keep the environment moist. Keep in mind the worms will only eat about the top inch of material, so don't fill your compost bin with too much material too fast.

STEP 8:
Use the liquid that collects in the bottom tub to water your garden. When you have enough castings, you can add them to your garden as well.

HIGH-TECH PLANTS

Plants are constantly working in ways that feed, protect, and support animal life. Throughout history, humans have also invented many uses for plants. Burning wood has always fueled human civilizations. Today, new technologies are used to make renewable biofuels from plant materials. These energy sources may be cleaner and better for the environment than other fuels.

Scientists and engineers have also learned to adjust the natural properties of plants in order to make them stronger, more productive, or more nutritious as food sources. They can do this by breeding plants in certain ways or by changing their genetic makeup. When genes are changed in a lab, the resulting plants are called genetically modified organisms (GMOs). New technologies will continue to shape the important relationship between plants and people.

MAKER MAGIC

Even the earliest humans used plants as medicines. Many of the medicines used and developed today are still plant based. As new plants are discovered, scientists look for properties that can help cure diseases.

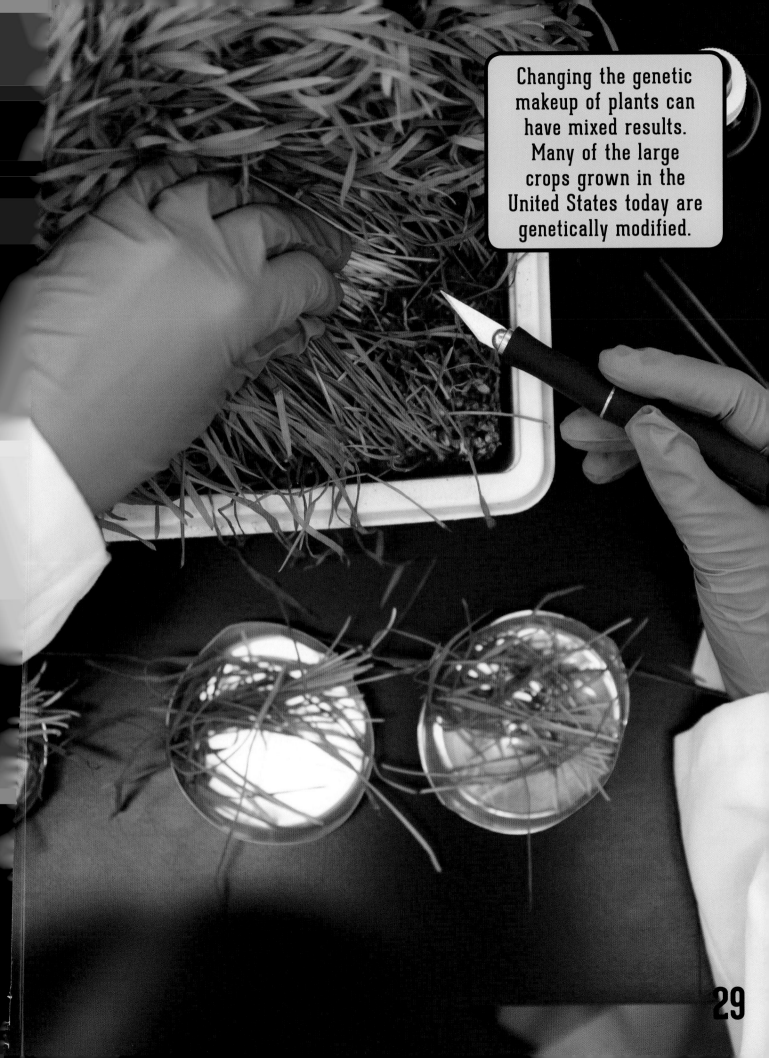

Changing the genetic makeup of plants can have mixed results. Many of the large crops grown in the United States today are genetically modified.

29

GOING GREEN

You don't have to look far to see examples of plant science at work around you. The nearest lawn, park, or garden is probably bursting with plant life. The world of plants is enormous, with hundreds of thousands of species. What can you observe happening in the plants around you? Use your questions about plant science to inspire new research and makerspace projects.

Take a hands-on approach and try creating a vertical garden in your yard or building a bee bath for thirsty pollinators. You can create a **terrarium** in your makerspace or experiment with a new technology to help your garden grow. Taking an active role in learning and sharing the experience with others will help you discover and understand the green neighbors who make Earth our home.

GLOSSARY

access: The ability to use or enter something.

category: A group of people or things similar in some way.

classify: To arrange in different classes.

define: To show or explain something completely.

describe: To tell someone what something is like.

evolve: To grow and change over time.

fibrous: Made of or containing fibers.

genetic: Referring to the parts of cells that control the appearance, growth, and other traits of living things.

genus: The scientific name for a group of plants or animals that share most features.

habitat: The natural home for plants, animals, and other living things.

nutrient: Something taken in by a plant or animal that helps it grow and stay healthy.

pigment: A natural coloring matter in plants.

technology: A method that uses science to solve problems and the tools used to solve those problems.

terrarium: An enclosed, transparent container for observing animals and plants indoors.

INDEX

B
binomial nomenclature, 10

C
carbon dioxide, 15
chlorophyll, 14

F
flowers, 13, 18, 20, 21
fruits, 13, 18

G
germination, 18

L
leaves, 6, 12, 13, 14, 15, 18, 25
light, 13, 14, 15

N
nutrients, 12, 24, 25, 26

O
oxygen, 14, 15, 24

P
photosynthesis, 14, 18, 24
phototropism, 13
pollen, 20

R
roots, 6, 12, 13, 15, 18, 24

S
seeds, 16, 17, 18, 19, 20
soil, 12, 15, 16, 17, 24, 26
stem, 6, 12, 13, 15, 18

W
water, 12, 13, 15, 16, 17, 18, 20, 24, 27

WEBSITES

Due to the changing nature of Internet links, PowerKids Press has developed an online list of websites related to the subject of this book. This site is updated regularly. Please use this link to access the list: www.powerkidslinks.com/stemmake/plantsci